DISCARDED

INSIDE THE HUMAN BODY

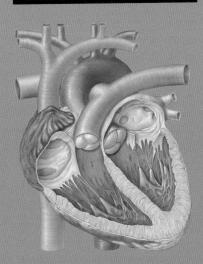

INTRODUCTION TO THE HUMAN BODY

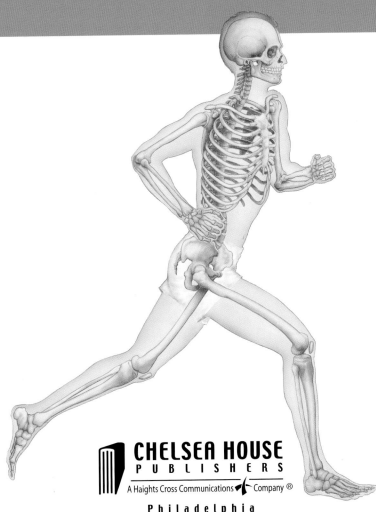

CHELSEA HOUSE PUBLISHERS
A Haights Cross Communications ® Company

Philadelphia

First hardcover library edition published
in the United States of America in
2006 by Chelsea House Publishers,
a subsidiary of Haights Cross Communications.
All rights reserved.

A Haights Cross Communications Company ®

www.chelseahouse.com

Library of Congress Cataloging-in-Publication
applied for.
ISBN 0-7910-9012-1

Project and realization
Parramón, Inc.

Texts
Adolfo Cassan

Translator
Patrick Clark

Graphic Design and Typesetting
Toni Inglés Studio

Illustrations
Marcel Socías Studio

First edition - September 2004

Printed in Spain
© Parramón Ediciones, S.A. – 2005
Ronda de Sant Pere, 5, 4ª planta
08010 Barcelona (España)
Norma Editorial Group

www.parramon.com

The whole or partial reproduction of this work by any means or procedure, including printing, photocopying, microfilm, digitalization, or any other system, without the express written permission of the publisher, is prohibited.

TABLE OF CONTENTS

4	**An Amazing Machine**
8	**The Circulatory System** Transporting Blood
10	**The Heart** The Motor of the Circulatory System
12	**The Respiratory System** Our Source of Oxygen
14	**The Nervous System** Our Central Computer
16	**The Digestive System** Getting Nutrients
18	**The Skin** Our Body's Shield
20	**The Urinary Tract** The Purification System
22	**The Reproductive System** The Male Genitals
24	**The Reproductive System** The Female Genitals
26	**The Endocrine System** Hormones: Chemical Messengers
28	**The Skeletal System** The Framework of Our Body
30–32	**The Muscular System:** Muscles: Strength and Movement. Index

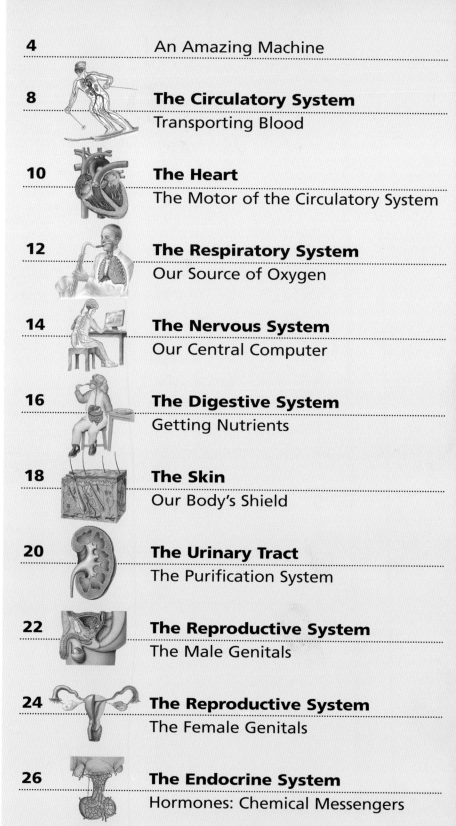

A PERFECT MACHINE

This book introduces readers to the wonders of the human body. Much like a well-designed machine, our bodies have special structures that play the role of the "parts" and "gears" of a machine.

This book examines the body's different systems that work together to keep it functioning properly. Each section has a large illustration that shows the structure of an organ or system, and gives some brief explanations about its main parts, its purpose, and the way it works.

This book is a practical, educational, and challenging guide to the human body.

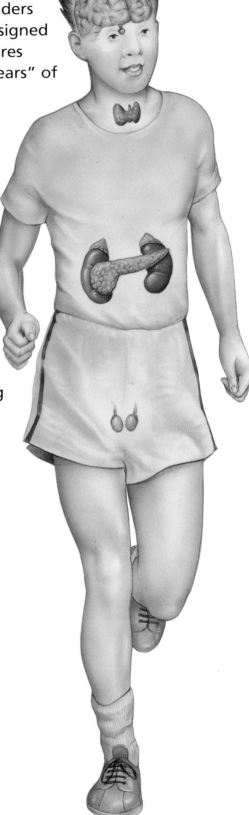

INTRODUCTION

AN AMAZING MACHINE

The human body is made up of cells—the basic units of life. There are about 200 billion tiny cells in our bodies!

Unlike the simplest living organisms, such as bacteria or protozoa that are made of a single cell, the human body is very complex. It is made up of 200 billion tiny cells! Although these units have the same basic parts, they differ in size, shape, and prescribed activity. The cells of the human body work together in perfect coordination.

The coordinated activities of the body's cells give each individual body its particular makeup and properties. Our bodies allow us to move, acquire nutrition, feel, interact with others, and reproduce. But how can microscopic cells come together to form an organism as complex as ours?

THE ORGANIZATION OF THE BODY

The different cells that make up the human body are highly organized. This means that they are not arranged in a random way. Instead, they are grouped according to the characteristics of each type of cell.

Sometimes, cells combine with substances such as mineral salts or fibers to form genuine tissues. There are many different kinds of tissues: epithelial, connective, muscle, and nerve. These tissues, in turn, combine to form all the structures and parts of the body.

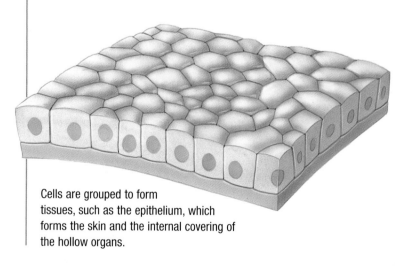

Cells are grouped to form tissues, such as the epithelium, which forms the skin and the internal covering of the hollow organs.

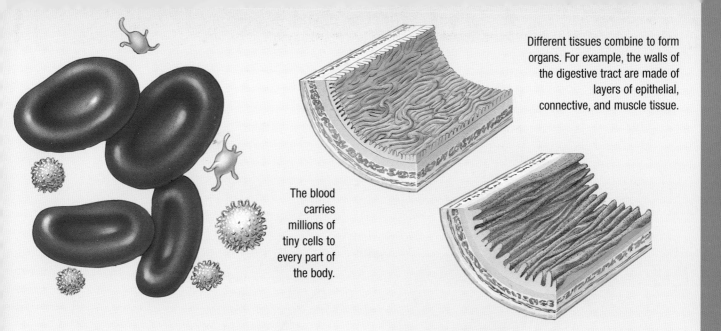

The blood carries millions of tiny cells to every part of the body.

Different tissues combine to form organs. For example, the walls of the digestive tract are made of layers of epithelial, connective, and muscle tissue.

The many structures and parts of the human body are specifically adapted to carry out precise functions. Our bodies have organs of every kind: solid, hollow, external, internal, large, small, and vital. There are even organs that are so special that they are made up mainly by a specific type of tissue. For example, skin is made up of epithelial tissue; bones contain osseous tissue; and nerves are made of nerve tissue.

BODY SYSTEMS

Some organs perform their functions on their own in an apparently automatic way. The skin, for example, protects our bodies and keeps us safe from the dangers of the outside world. Of course, this is a simple way to look at it, since the skin also depends on other organs for the nutrition it needs to carry out its activity. Most of our organs need to work together to form a functional unit or body system.

The mouth, the stomach, and the bladder work together to form the digestive system. The word *system* is used to describe functional units of organs made up of basically the same kind of tissue, as in the case of the nervous system, or by structures that are not anatomically related but work in the same way, like the various glands of the endocrine system.

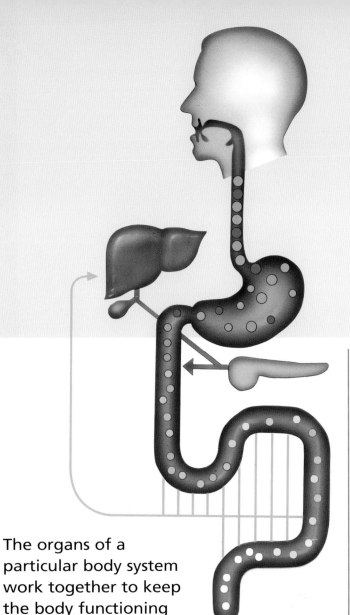

Combinations of different organs that have the same mission make up body systems. An example of this is the digestive system, which breaks up and processes food.

The organs of a particular body system work together to keep the body functioning properly. For example, some organs take in, process, and distribute needed substances such as oxygen or nutrients throughout the body. Others regulate the physical/chemical makeup of the body and keep it properly balanced. Some protect us from or remove toxic substances, and others are involved in overall control, sustenance and movement, or reproduction.

A FUNCTIONAL UNIT

To keep us alive and in good health, the various tissues and organs have to maintain perfectly coordinated activity, since they are largely dependent on each other.

Some organs may be considered more important than others, because they play a role in exchanging matter and energy with the world around us, or because they handle functions that we need to stay alive. But this is not really the case. As in society, where each person has a role to play, all the parts of our body have important tasks. The respiratory system, which lets us take in oxygen, is vital, as is the digestive system, which brings us nutrients from food. The same can be said of the circulatory system, which moves the blood that carries

The nervous system controls and regulates the function of many organs, such as the secretion of digestive juices by the stomach.

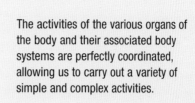

The activities of the various organs of the body and their associated body systems are perfectly coordinated, allowing us to carry out a variety of simple and complex activities.

oxygen and nutrients throughout the body, or of the kidneys, which cleanse the blood and get rid of waste products through urination. This is also true of the skin, which acts as a protective barrier and helps maintain body temperature, and the nervous and endocrine systems, which control and regulate the functioning of the body as a whole.

A VISION

It has taken a great deal of time and effort over the course of history to get a clear idea of the makeup and function of the human body. In the 2nd century A.D., Greek doctor Galen of Pergamon laid the foundation for a new era of knowledge about the human body. Since then, many scientists have contributed to our understanding of the complex processes of the human body.

To understand the anatomy and physiology of the human body, it is important to put together basic ideas about each part and the whole.

Galen of Pergamon, a Greek doctor in the 2nd century A.D., laid out the basis for our present understanding of human anatomy and physiology.

THE CIRCULATORY SYSTEM

TRANSPORTING BLOOD

The circulatory system is made up of a complex network of vessels that, with the aid of the constant pumping of the heart, continually move blood throughout the body. The blood brings oxygen and nutrients to the tissues, collects waste products, and distributes these waste products to the organs that will remove them from the body.

subclavical vein ■
drains blood from the upper limbs

superior vena cava ■
collects the venous blood from the upper part of the body and carries it to the heart

heart ■
central organ of the circulatory system; pumps blood rhythmically to the arteries so it can be carried throughout the body

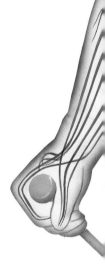

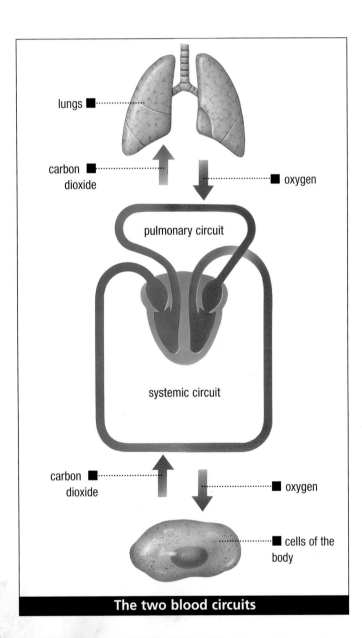

The two blood circuits

ARTERIES AND VEINS

In our body, there are two types of circulatory vessels that perform complementary functions: arteries, which are ducts with elastic walls that carry the blood pumped by the heart throughout the body; and veins, which are more relaxed and carry the blood back to the heart. There are many arteries and veins found throughout the body, but only the main ones have individual names.

The human body has two circulatory circuits that work together. One is the systemic circuit: Oxygen-rich blood is sent from the heart to the aorta, travels through many different arteries to all the tissues, and then returns to the heart, low in oxygen and loaded with carbon dioxide, through the veins. The other is the pulmonary circuit: Venous blood is pushed by the heart to the pulmonary artery, picks up oxygen in the lungs, and returns to the heart, the point of departure for the systemic circuit.

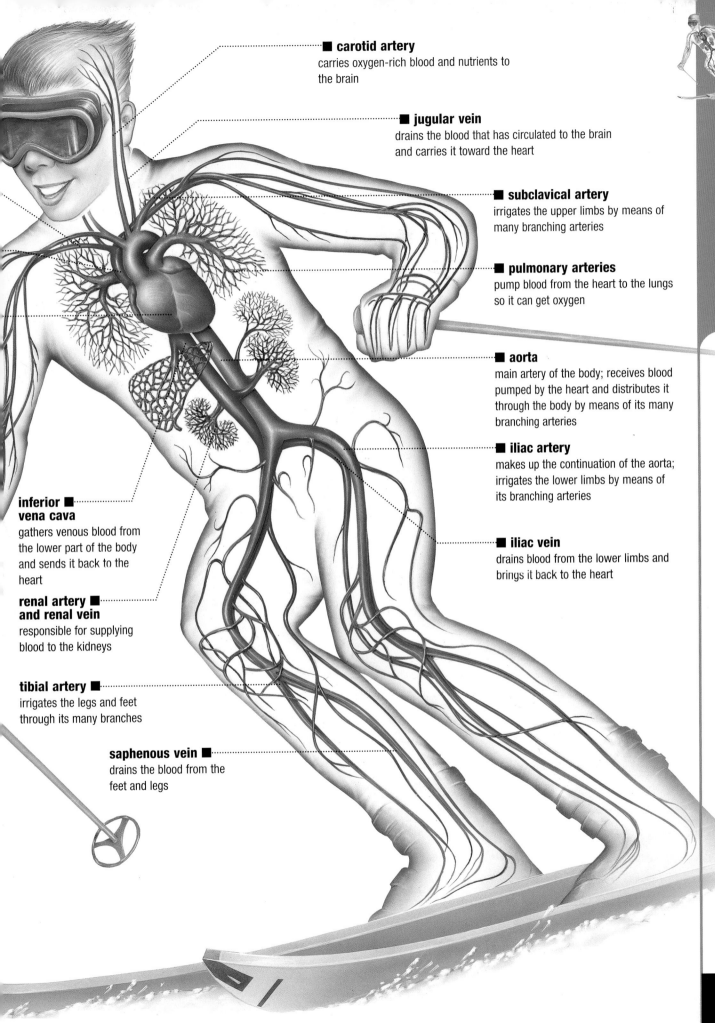

THE CIRCULATORY SYSTEM

- **carotid artery**
  carries oxygen-rich blood and nutrients to the brain

- **jugular vein**
  drains the blood that has circulated to the brain and carries it toward the heart

- **subclavical artery**
  irrigates the upper limbs by means of many branching arteries

- **pulmonary arteries**
  pump blood from the heart to the lungs so it can get oxygen

- **aorta**
  main artery of the body; receives blood pumped by the heart and distributes it through the body by means of its many branching arteries

- **iliac artery**
  makes up the continuation of the aorta; irrigates the lower limbs by means of its branching arteries

- **iliac vein**
  drains blood from the lower limbs and brings it back to the heart

- **inferior vena cava**
  gathers venous blood from the lower part of the body and sends it back to the heart

- **renal artery and renal vein**
  responsible for supplying blood to the kidneys

- **tibial artery**
  irrigates the legs and feet through its many branches

- **saphenous vein**
  drains the blood from the feet and legs

8
9

THE HEART

THE MOTOR
OF THE CIRCULATORY SYSTEM

The heart is a hollow organ with muscular walls, and is divided into four chambers with four valves that control the direction of blood flow. With each beat, the heart pumps oxygen-rich blood toward the arteries, then fills with oxygen-poor blood that comes through the veins. After the blood picks up oxygen in the lungs, the heart pushes it back to the arteries so it can be carried throughout the body, in a cycle that repeats all our lives.

vena cava ■
vessel that carries oxygen-poor blood that has already circulated through the body back to the heart

right atrium ■
chamber of the heart that receives the oxygen-poor blood that has already circulated through the body and pushes it to the right ventricle

right ventricle ■
chamber of the heart that receives oxygen-poor blood from the right atrium and pushes it toward the lungs

AN UNENDING BEAT
The heart beats without interruption from before we are born until we die. It is estimated that, over the course of an 80-year lifespan, the heart beats no fewer than 3 billion times!

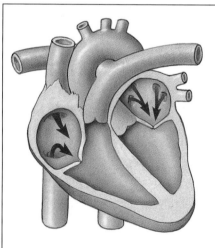

DIASTOLE
The atriums dilate (open) and are filled with blood coming from the veins

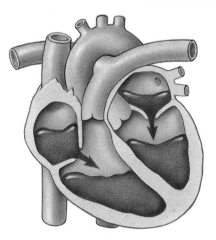

ATRIAL SYSTOLE
The atriums contract and pump blood to the ventricles

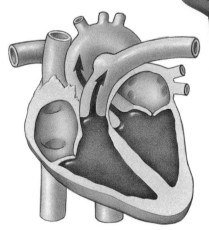

VENTRICULAR SYSTOLE
The ventricles contract and pump blood to the arteries

The cardiac cycle

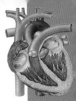

■ **aorta**
vessel that receives oxygen-rich blood from the heart and distributes it by means of its tributaries throughout the body

■ **pulmonary artery**
vessel that carries oxygen-poor blood to the lungs

■ **pulmonary veins**
vessels that carry oxygen-rich blood from the lungs to the heart

■ **left atrium**
chamber of the heart that receives oxygen-rich blood coming from the lungs and pushes it into the left ventricle

■ **atrioventricular valves**
valves that allow blood to move from each atrium to the ventricle of each side, and keep it from going back

■ **left ventricle**
chamber of the heart that receives oxygen-rich blood from the left atrium and pushes it to the arteries so it can be distributed all over the body

■ **myocardium**
thick muscle layer of the heart wall

THE HEART

10
11

THE RESPIRATORY SYSTEM

OUR SOURCE OF OXYGEN

The respiratory system has a vital function: It exchanges gases between the blood and the air around us. Every time we breathe, air goes into the lungs and lets oxygen into the blood. The heart then pumps this blood to the pulmonary arteries. At the same time, the blood lets out carbon dioxide to the air as a waste product, so that it can be removed from the body when we exhale. Purified blood then returns to the heart.

mouth ■
secondary pathway for the entry and exit of air, although it is usually considered part of the digestive system

trachea ■
duct that connects the larynx with the left and right main bronchi

bronchi ■ (left main, right main)
ducts that start at the trachea and branch out into smaller ducts that are embedded in the pulmonary tissue

- pulmonary vein
- bronchiole
- pulmonary artery
- alveoli

The pulmonary unit

Pulmonary tissue is made up of millions of tiny hollow sacks filled with air, the alveoli, which are surrounded by small circulatory vessels. The exchange of gases between the blood and the air occurs through the thin walls of these vessels.

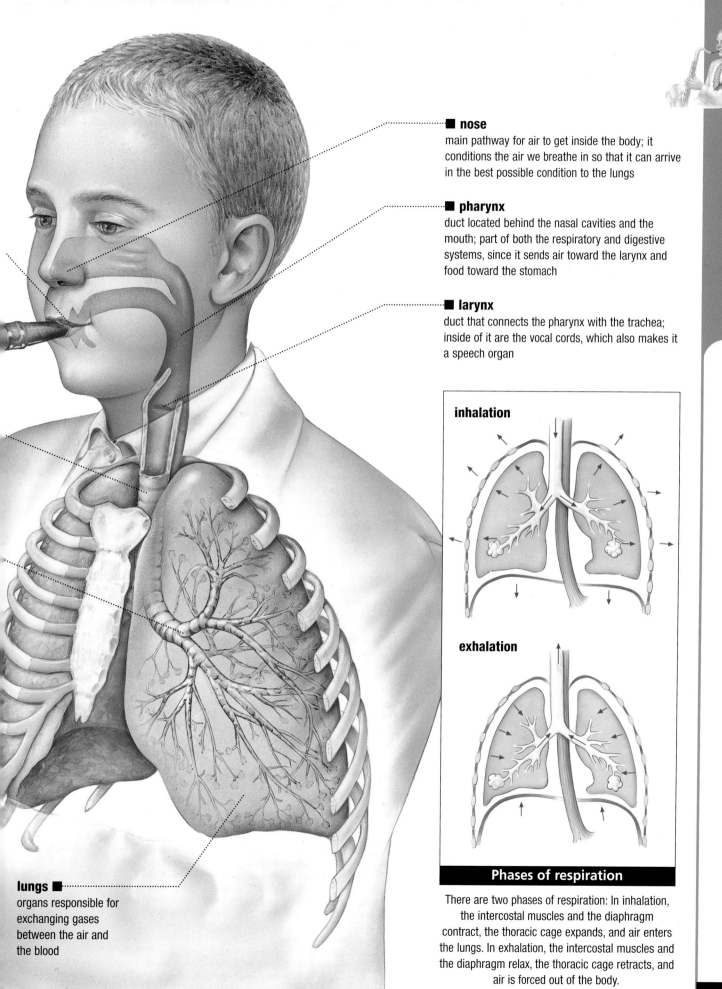

nose
main pathway for air to get inside the body; it conditions the air we breathe in so that it can arrive in the best possible condition to the lungs

pharynx
duct located behind the nasal cavities and the mouth; part of both the respiratory and digestive systems, since it sends air toward the larynx and food toward the stomach

larynx
duct that connects the pharynx with the trachea; inside of it are the vocal cords, which also makes it a speech organ

lungs
organs responsible for exchanging gases between the air and the blood

Phases of respiration

There are two phases of respiration: In inhalation, the intercostal muscles and the diaphragm contract, the thoracic cage expands, and air enters the lungs. In exhalation, the intercostal muscles and the diaphragm relax, the thoracic cage retracts, and air is forced out of the body.

THE NERVOUS SYSTEM

OUR CENTRAL COMPUTER

The nervous system coordinates all the functions of our body. The nervous system's main organ is the brain, which controls our conscious, voluntary actions, and also regulates the unconscious, automatic activity of our internal organs. It is also responsible for the complex interactions our bodies have with the outside world, and is where intellectual activity occurs.

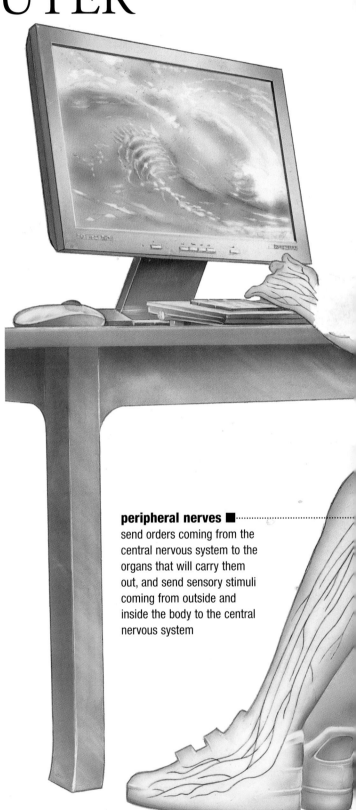

peripheral nerves ■
send orders coming from the central nervous system to the organs that will carry them out, and send sensory stimuli coming from outside and inside the body to the central nervous system

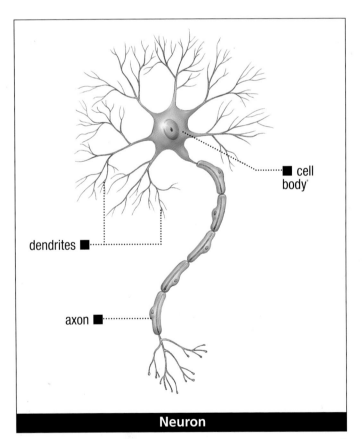

Neuron

All the structures of the nervous system are formed by a special type of cell called the neuron, which is made up of a cell body and two kinds of extensions: shorter dendrites and longer axons.

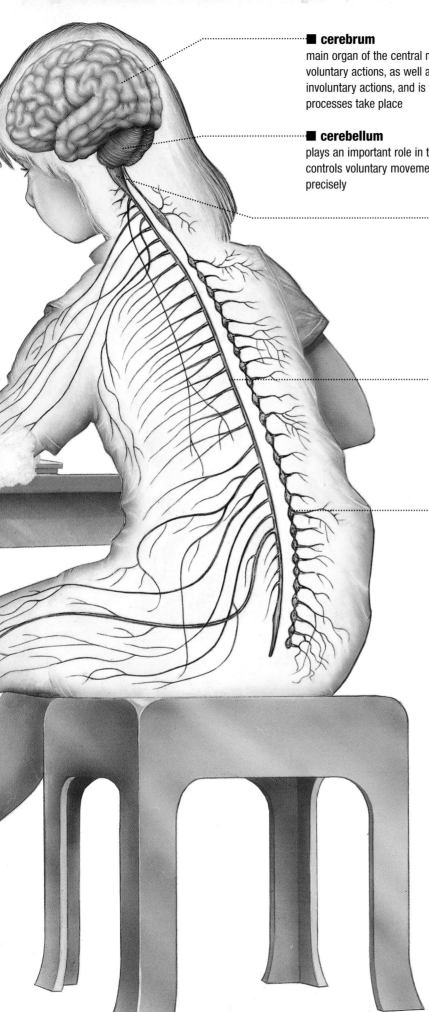

■ **cerebrum**
main organ of the central nervous system; controls all voluntary actions, as well as most of the body's involuntary actions, and is where our mental processes take place

■ **cerebellum**
plays an important role in the control of balance, and controls voluntary movements so they can be carried out precisely

■ **encephalic trunk**
made of the brain stem, the spinal protuberance, and the spinal bulb, this is a connection pathway from the cerebrum and cerebellum to the spinal medulla. It houses the nerve centers that control vital functions such as breathing and heartbeat

■ **spinal medulla**
connects the upper nerve centers with the body, since this is where all the nerves start, both those that send orders to body organs and those that relay messages about stimuli

■ **autonomic nervous system**
regulates automatic bodily functions, such as keeping the body temperature steady and controlling breathing and digestion

IS THE BRAIN A GIANT COMPUTER?

The brain is made up of millions of circuits, somewhat like microchips, that allow it to analyze information and react instantly with the most appropriate response. But it is much more than a computer, because it carries out a series of tasks that no computer can do: It allows us to think, remember, imagine, feel, and perceive.

THE DIGESTIVE SYSTEM

GETTING NUTRIENTS

The digestive system has a very important mission: It breaks down food into nutrients so that the body can absorb them. The blood then carries these nutrients to all the corners of the body, and uses them to provide the energy needed to form tissues and perform important functions.

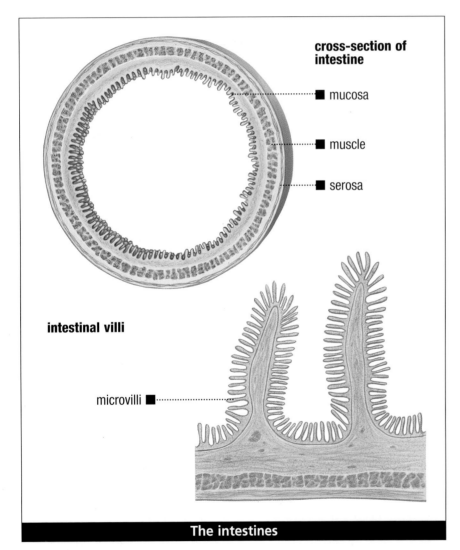

cross-section of intestine
- mucosa
- muscle
- serosa

intestinal villi

microvilli

The intestines

The interior wall of the intestine is covered with a mucus layer that has a large number of projections, called villi. Each of the villi has many hair-like formations, the microvilli, which make the surface that comes in contact with food much larger.

bile duct
stores bile made in the liver; after eating, dumps food to the duodenum

duodenum
first section of the small intestine, where food is broken down by the action of intestinal enzymes, pancreatic fluid, and bile, thereby freeing the basic nutrients

pancreas
secretes pancreatic fluid, which is made of enzymes needed for the digestion of food

large intestine
made up of the colon and rectum; it is the last part of the digestive tube

FOOD'S PATH

Food stays in the stomach for 2 to 4 hours, travels through the small intestine for 3 to 4 hours, and then whatever is left moves through the large intestine for 10 to 48 hours before being eliminated through defecation.

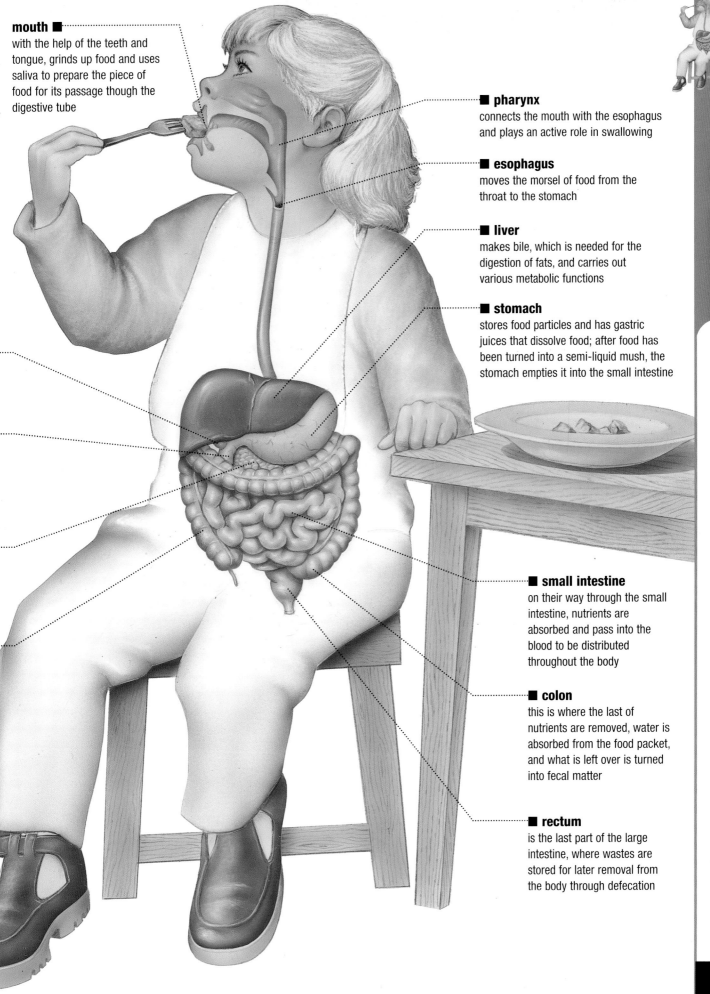

mouth
with the help of the teeth and tongue, grinds up food and uses saliva to prepare the piece of food for its passage though the digestive tube

pharynx
connects the mouth with the esophagus and plays an active role in swallowing

esophagus
moves the morsel of food from the throat to the stomach

liver
makes bile, which is needed for the digestion of fats, and carries out various metabolic functions

stomach
stores food particles and has gastric juices that dissolve food; after food has been turned into a semi-liquid mush, the stomach empties it into the small intestine

small intestine
on their way through the small intestine, nutrients are absorbed and pass into the blood to be distributed throughout the body

colon
this is where the last of nutrients are removed, water is absorbed from the food packet, and what is left over is turned into fecal matter

rectum
is the last part of the large intestine, where wastes are stored for later removal from the body through defecation

THE DIGESTIVE SYSTEM

16
17

THE SKIN

OUR BODY'S SHIELD

The skin covers the human body. It helps determine our appearance, and also carries out various functions. Skin acts as a protective barrier against many kinds of external dangers. It helps maintain constant body temperature and internal chemistry, and has an energy reserve in its adipose (fatty) tissue. Skin also acts as a sensory organ, providing us with the sense of touch.

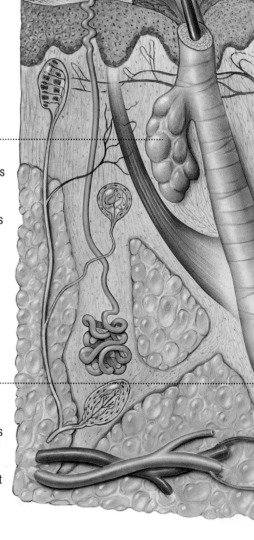

sebaceous glands ■ make a fatty secretion that forms a protective film over the epidermis and keeps the hairs moist

hair follicles ■ formations where hair is generated. These are filaments of lesser or greater thickness and different colors that are found over most of the skin's surface

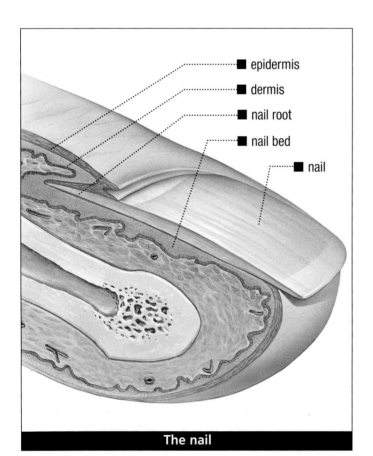

- epidermis
- dermis
- nail root
- nail bed
- nail

The nail

The nail is a cutaneous (formed of skin) extension. It is a plate that grows out of the epidermis, at the rate of about one millimeter per day. It is made up mainly of a hard protein called keratin, which also forms hair.

THE LARGEST ORGAN

Although we don't often think of it this way, the skin is a real body organ, since it carries out very specific functions. It is actually the largest organ in the body: In an adult, it extends over 15–20 square feet (1.4–1.9 square meters). The epidermis and dermis alone weigh around 9 pounds (4 kg).

sensory receptors ■
nerve endings and more complex structures whose mission is to detect pressure, tactile (touch), and thermal (temperature) sensations

■ **sweat glands**
produce sweat, a secretion made of water and small amounts of salty minerals and waste products that help regulate body temperature, since they have a cooling effect as they evaporate

■ **epidermis**
surface layer of the skin, formed by layers of cells that are in direct contact with the outside world

■ **dermis**
middle layer of the skin, made of cells and fibers of connective tissue, that houses many sensory receptors

■ **hypodermis**
the deepest layer of the skin, which has different thickness in different parts of the body; contains a lot of adipose (fatty) tissue, which acts as an energy reserve and a thermal insulator

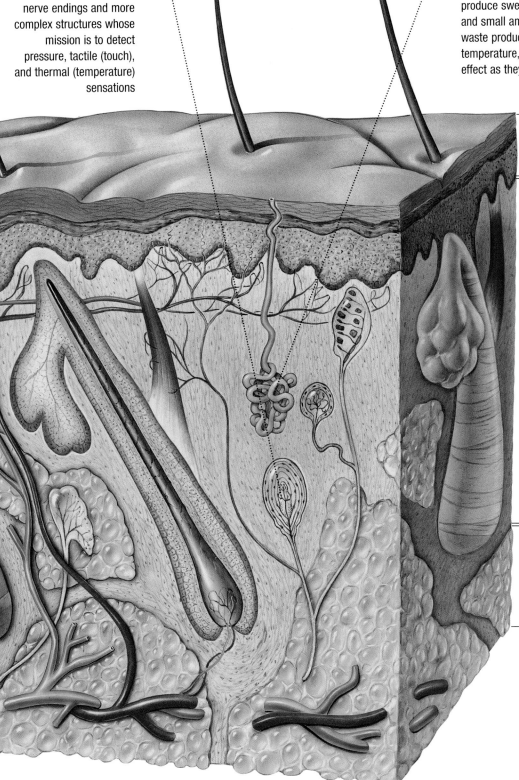

THE URINARY TRACT

THE PURIFICATION SYSTEM

The urinary tract filters the blood that circulates through our body. It regulates the blood's composition and, by means of the urine that the kidneys continually make, removes wastes like excess water, salty minerals, toxic substances, and metabolic residue that could harm the body in large quantities. Essentially, the urinary tract is a purification system that we need to stay alive.

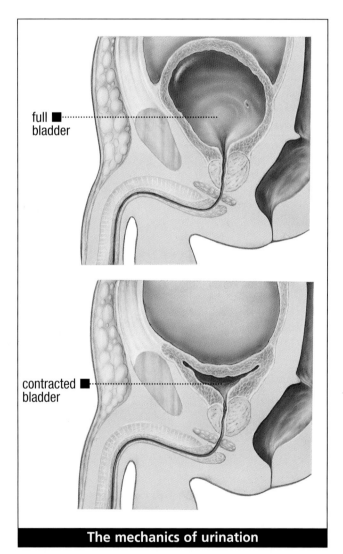

The mechanics of urination

The bladder has a limited amount of space to store the urine that is continually produced by the kidneys. When the bladder is full, we feel the desire to urinate. When urination takes place, a valve that connects the bladder with the urethra opens, the walls of the bladder contract, and urine is forced out of the body.

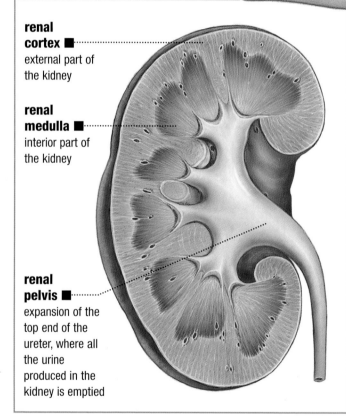

renal cortex external part of the kidney

renal medulla interior part of the kidney

renal pelvis expansion of the top end of the ureter, where all the urine produced in the kidney is emptied

URINE

How much urine the kidneys make depends on different factors—most of all how much liquid we take in, both through food and drink. Even if we drink and eat very little, however, the kidneys have to make a certain amount of urine to filter the waste products of the blood. Under normal circumstances, they produce between a quart and half a gallon of urine every day.

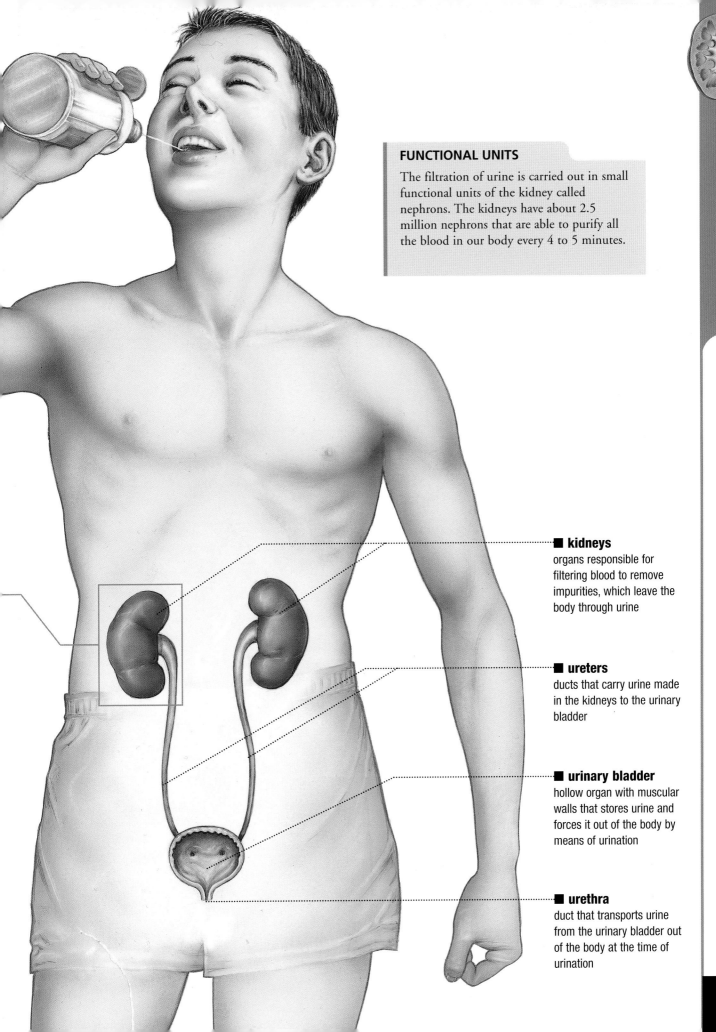

FUNCTIONAL UNITS

The filtration of urine is carried out in small functional units of the kidney called nephrons. The kidneys have about 2.5 million nephrons that are able to purify all the blood in our body every 4 to 5 minutes.

THE URINARY TRACT

■ **kidneys**
organs responsible for filtering blood to remove impurities, which leave the body through urine

■ **ureters**
ducts that carry urine made in the kidneys to the urinary bladder

■ **urinary bladder**
hollow organ with muscular walls that stores urine and forces it out of the body by means of urination

■ **urethra**
duct that transports urine from the urinary bladder out of the body at the time of urination

THE REPRODUCTIVE SYSTEM

THE MALE GENITALS

Male genitals consist of a set of organs designed to carry out sexual activity. These organs participate in the process of reproduction. The main goal of the organs is to carry male gametes, called sperm cells, into the female body, where they may fertilize a female egg cell that will grow into a fetus.

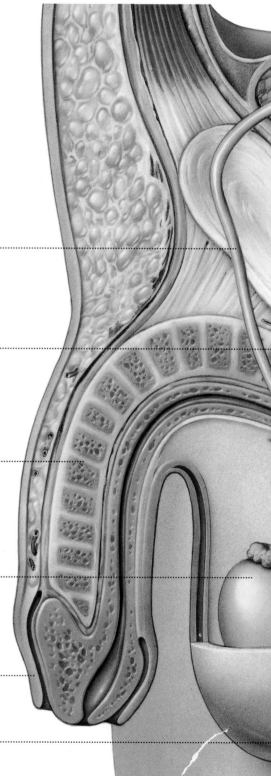

vas deferens ■
tube by which mature sperm cells travel from the epididymis to the ejaculatory duct on the way out of the body

ejaculatory duct ■
tube that sends the sperm from the vas deferens and the secretions of the seminal vesicle toward the urethra

penis ■
organ that carries out the sex act

testicle ■
male gonad that produces sperm cells and also makes the male sex hormones

foreskin ■
fold of skin that covers the tip of the penis

scrotum ■
cutaneous sack that hangs at the base of the penis and holds the testicles

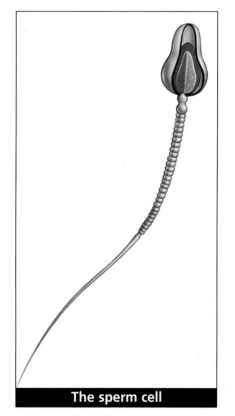

The sperm cell

Sperm cells are the male gametes, tiny cells with a head and a long tail that float and move in semen. Their purpose is to enter the female genitals in search of an egg cell to fertilize.

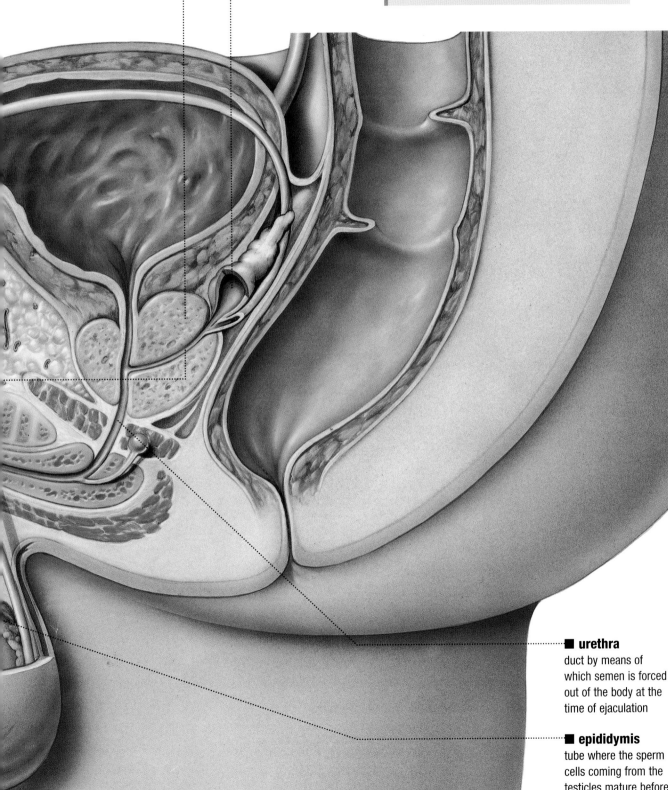

prostate
gland that produces a secretion that contains nutrients for the sperm cells that will make up part of the semen

seminal vesicle
gland that produces a secretion that carries the sperm cells and provides them with nutrients

WORKING NONSTOP
Inside the testicles, there is frantic activity, because the production of sperm cells goes on all the time. In each ejaculation, there are between 2 and 6 milliliters of semen, and the concentration of sperm cells is between 20 and 90 million per milliliter.

THE REPRODUCTIVE SYSTEM

urethra
duct by means of which semen is forced out of the body at the time of ejaculation

epididymis
tube where the sperm cells coming from the testicles mature before following their path toward the outside

THE REPRODUCTIVE SYSTEM

THE FEMALE GENITALS

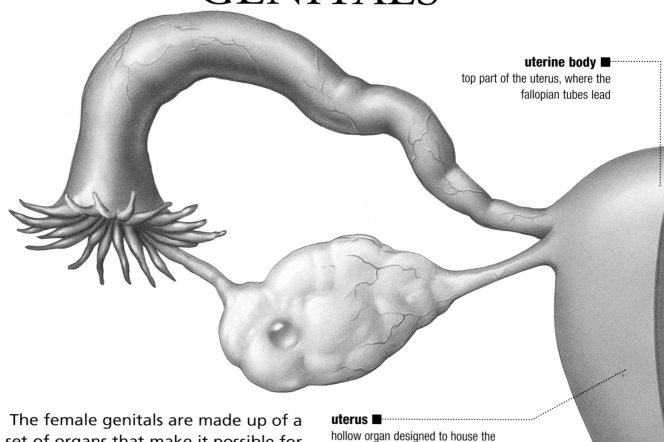

The female genitals are made up of a set of organs that make it possible for women to carry out sexual activity. They are specially arranged to play a role in the process of reproduction, since they are able to house the united egg and sperm cell and to provide everything it needs to develop during pregnancy.

uterine body ■
top part of the uterus, where the fallopian tubes lead

uterus ■
hollow organ designed to house the fertilized egg and shelter the fetus during pregnancy

uterine cavity ■
central cavity of the uterus, connected with the vagina through the canal of the cervix

vagina ■
duct with elastic walls that is connected with the outside world; designed to receive the penis and deliver the fetus

EGG SUPPLY

When a girl is born, the ovaries contain 400,000 immature eggs, but only a few hundred will periodically mature throughout a woman's life. The fertile stage begins at puberty, when a girl begins to menstruate, and continues until menopause, when the menstrual cycle stops.

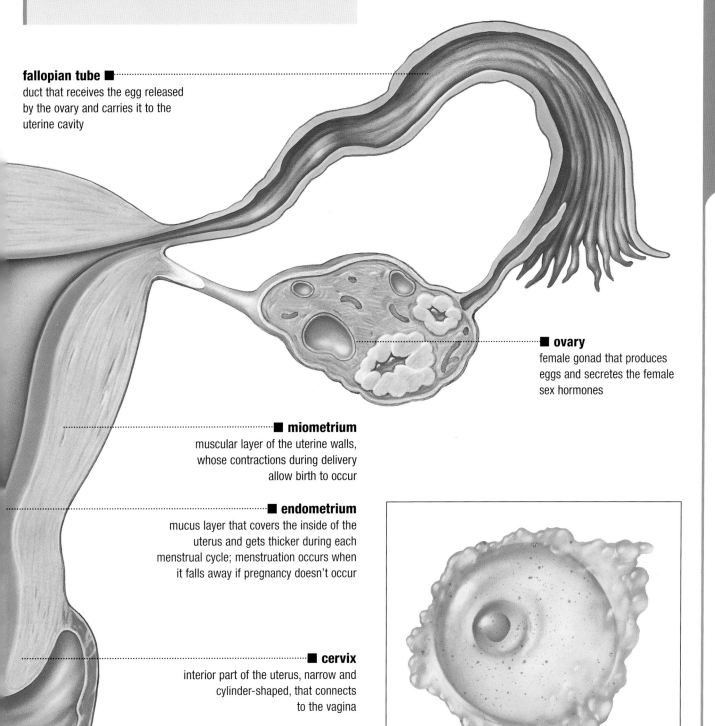

fallopian tube ■
duct that receives the egg released by the ovary and carries it to the uterine cavity

■ **ovary**
female gonad that produces eggs and secretes the female sex hormones

■ **miometrium**
muscular layer of the uterine walls, whose contractions during delivery allow birth to occur

■ **endometrium**
mucus layer that covers the inside of the uterus and gets thicker during each menstrual cycle; menstruation occurs when it falls away if pregnancy doesn't occur

■ **cervix**
interior part of the uterus, narrow and cylinder-shaped, that connects to the vagina

The egg cell

This is the female gamete. After being fertilized by the male gamete (the sperm cell), it forms a zygote.

THE ENDOCRINE SYSTEM

HORMONES: CHEMICAL MESSENGERS

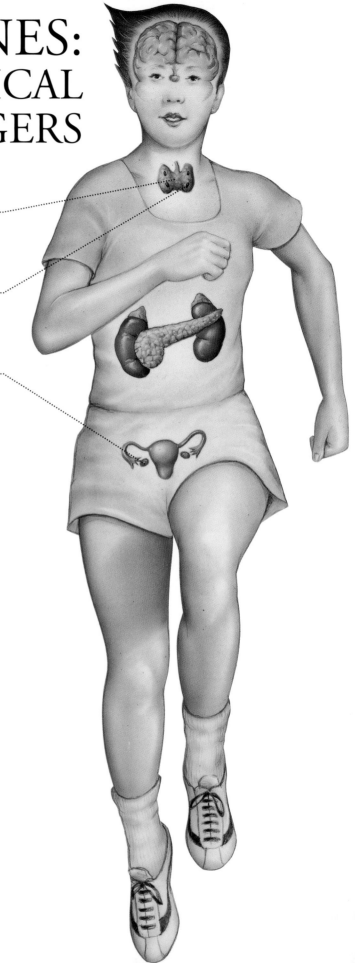

thyroid ■
gland that produces hormones that stimulate the body's metabolism; very important for physical and mental growth in children

parathyroids ■
glands that produce hormones that help regulate calcium and phosphorus levels in the blood

gonads ■
glands that secrete hormones responsible for the development of secondary sex characteristics, and where the gametes or germ cells mature. In men, these are known as the testicles; in women, they are called the ovaries

The endocrine system is made up of a set of internal secretion glands that release hormones into the blood. Hormones are chemical messengers that travel through the bloodstream to all parts of the body and control the function of different tissues and organs and regulate metabolism, growth, and physical development.

FUNCTIONS OF THE HYPOTHALAMUS
Besides producing hormones that control the activity of the pituitary gland (and, thereby, the endocrine system as a whole), the hypothalamus of the brain also regulates thirst, appetite, body temperature, and sleep.

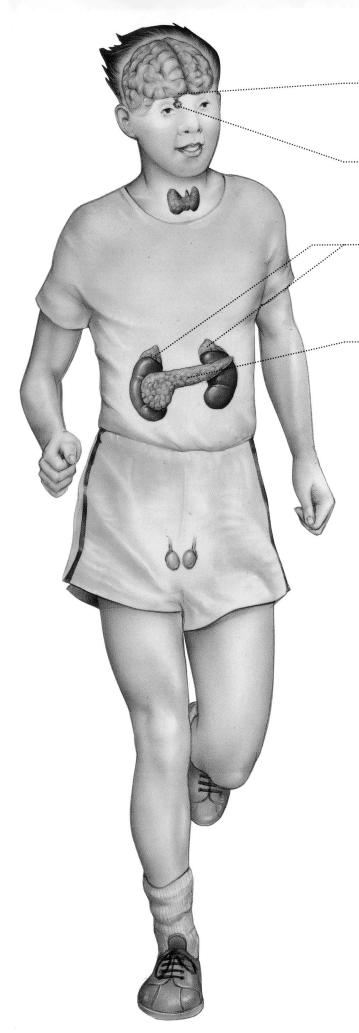

■ **hypothalamus**
structure that acts as a link between the nervous system and the endocrine system, and controls the activity of the pituitary gland, coordinating the activity of the whole endocrine system

■ **pituitary gland**
gland that produces hormones that regulate certain tissues and controls the activity of other glands of the endocrine system

■ **suprarenal glands**
glands that produce various hormones. Some control nutrient metabolism and the water/salt balance of the body, while others play a role in the autonomic nervous system

■ **pancreas**
gland that secretes two hormones, insulin and glucagon, that regulate glucose metabolism and blood concentration

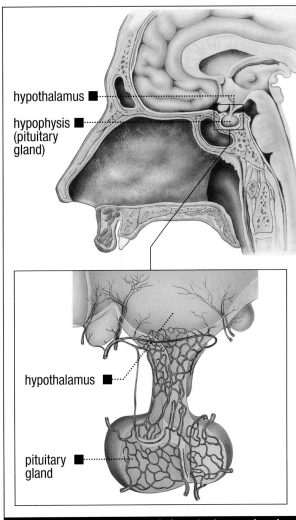

The hypothalamus and the pituitary gland

A dense network of circulatory vessels connects the hypothalamus and the pituitary gland. The hypothalamus controls every endocrine gland in the body.

THE SKELETAL SYSTEM

THE FRAMEWORK OF OUR BODY

The skeletal system is responsible for protection and mobility. Joints connect our bones to our skeletal muscles, which results in the ability to move our bodies. Basically, the bones act as lever arms, the joints serve as support points, and the muscles give us the strength we need to produce movements.

spinal column ■
formed by stacked vertebrae, makes up the axis of our skeleton. The spinal column is relatively mobile, allowing us to bend forward and sideways

humerus ■
is the arm bone, thick and very resistant

ribs ■
12 pairs of ribs form the thoracic cage and help protect fragile organs such as the lungs and heart

radius ■
ulna ■
are the bones of the forearm, the first located outside and the second inside when the arms are hanging and the palm of the hand is open, facing forward

iliac bone ■
flat bone that forms the pelvis, a bony ring that transfers the weight of the body to the lower limbs

femur ■
is the thigh bone—the longest, thickest, and strongest bone in the whole skeleton

- phalanges ■
- metacarpals ■
- wrist bones ■
- radius ■
- ulna ■

The wrist and the hand

The wrist is formed by eight different bones and makes up the joint that allows us to flex and extend the hand. The hand has five metacarpal bones that correspond to the fingers, each of which is made up of three small bones called phalanges. The thumb, which is thicker and shorter, has only two phalanges.

MORE THAN 200 BONES

The human skeleton consists of 206 bones, each with its own name, although there are people who have an additional tiny bone between the joints of the fingers.

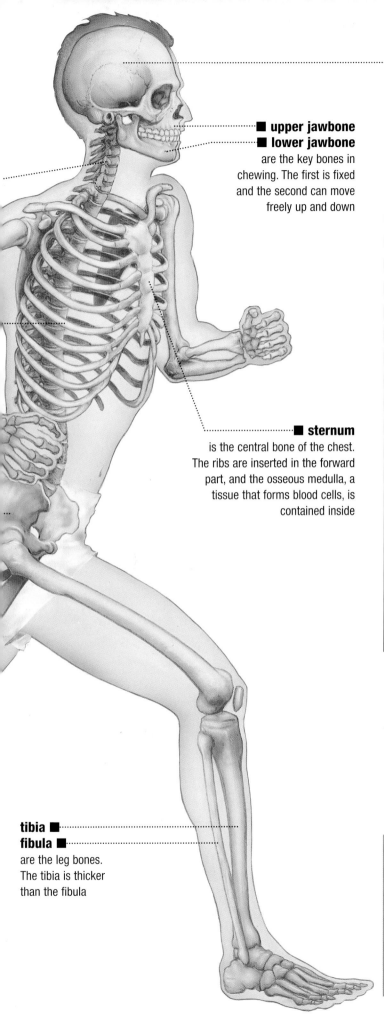

■ **cranium**
is a kind of natural "helmet" that protects the delicate brain inside, to prevent blows to the head from damaging it

■ **upper jawbone**
■ **lower jawbone**
are the key bones in chewing. The first is fixed and the second can move freely up and down

■ **sternum**
is the central bone of the chest. The ribs are inserted in the forward part, and the osseous medulla, a tissue that forms blood cells, is contained inside

tibia ■
fibula ■
are the leg bones. The tibia is thicker than the fibula

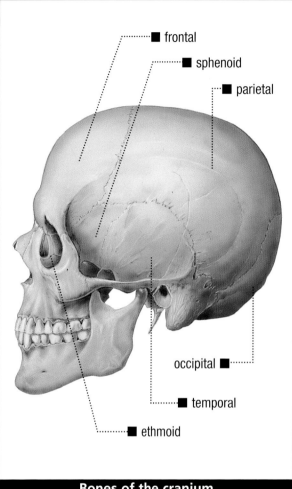

- frontal
- sphenoid
- parietal
- occipital
- temporal
- ethmoid

Bones of the cranium

The cranium is formed by eight bones: the frontal, two parietals, two temporals, the occipital, the sphenoid, and the ethmoid. At birth, some of these bones are not yet united. The cranium therefore has a certain degree of flexibility, but the bones join together soon after birth to create a rigid and compact structure.

BONE TISSUE

Bones are made of a special tissue that makes use of its own cells to create a blend of organic substances where minerals such as calcium and phosphorous are deposited. These minerals provide bone with its characteristic hardness and resistance. Although they appear to be inert pieces, bones are formed by living osseous tissue, tissue that is in a constant process of activity and renovation.

THE MUSCULAR SYSTEM

MUSCLES: STRENGTH AND MOVEMENT

The skeletal muscles, which are attached to the bones, form part of the muscular system. Due to their special ability to contract and relax, they can get shorter or longer. By changing their length, muscles can change the position of the bones to which they are attached. This is what allows us to have a full range of motion, from winking an eye to walking across the street.

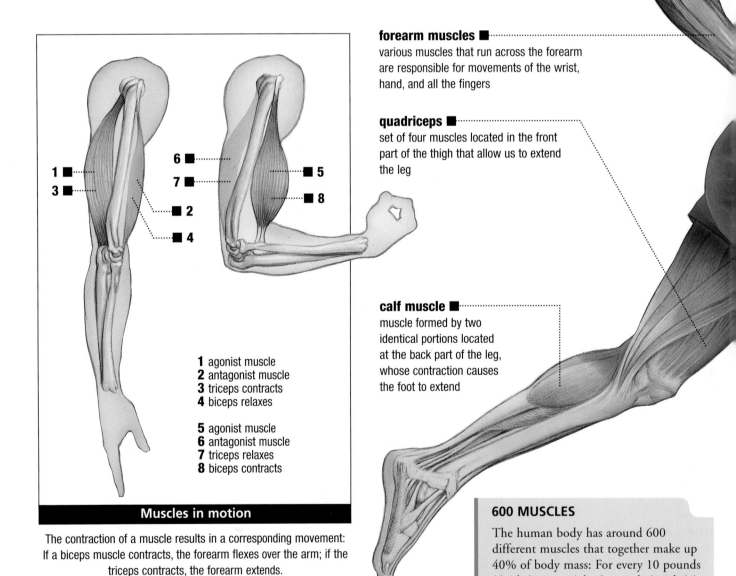

forearm muscles ■
various muscles that run across the forearm are responsible for movements of the wrist, hand, and all the fingers

quadriceps ■
set of four muscles located in the front part of the thigh that allow us to extend the leg

calf muscle ■
muscle formed by two identical portions located at the back part of the leg, whose contraction causes the foot to extend

1 agonist muscle
2 antagonist muscle
3 triceps contracts
4 biceps relaxes

5 agonist muscle
6 antagonist muscle
7 triceps relaxes
8 biceps contracts

Muscles in motion

The contraction of a muscle results in a corresponding movement: If a biceps muscle contracts, the forearm flexes over the arm; if the triceps contracts, the forearm extends.

600 MUSCLES

The human body has around 600 different muscles that together make up 40% of body mass: For every 10 pounds (4.5 kg) you weigh, 4 pounds (1.8 kg) is muscle.

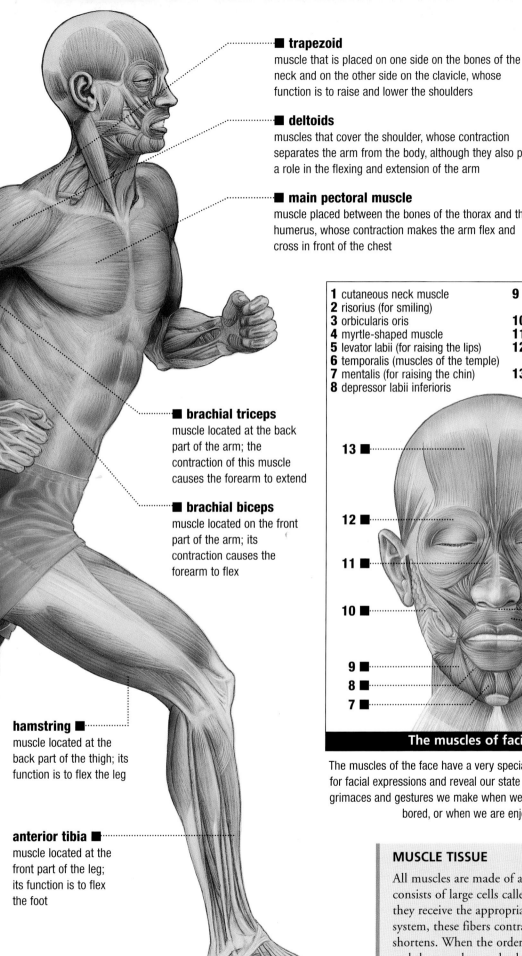

- **trapezoid**
  muscle that is placed on one side on the bones of the neck and on the other side on the clavicle, whose function is to raise and lower the shoulders

- **deltoids**
  muscles that cover the shoulder, whose contraction separates the arm from the body, although they also play a role in the flexing and extension of the arm

- **main pectoral muscle**
  muscle placed between the bones of the thorax and the humerus, whose contraction makes the arm flex and cross in front of the chest

- **brachial triceps**
  muscle located at the back part of the arm; the contraction of this muscle causes the forearm to extend

- **brachial biceps**
  muscle located on the front part of the arm; its contraction causes the forearm to flex

- **hamstring**
  muscle located at the back part of the thigh; its function is to flex the leg

- **anterior tibia**
  muscle located at the front part of the leg; its function is to flex the foot

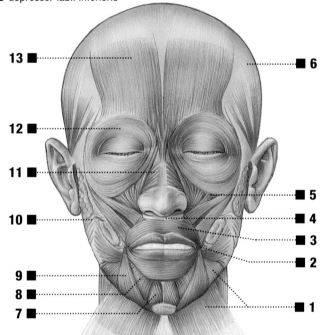

1 cutaneous neck muscle
2 risorius (for smiling)
3 orbicularis oris
4 myrtle-shaped muscle
5 levator labii (for raising the lips)
6 temporalis (muscles of the temple)
7 mentalis (for raising the chin)
8 depressor labii inferioris
9 depressor anguli oris (for frowning)
10 masseter (for chewing)
11 nasalis
12 orbicularis oculi (for closing the eyes)
13 frontalis (for raising eyebrows)

The muscles of facial gestures

The muscles of the face have a very special function. They are responsible for facial expressions and reveal our state of mind, since they produce the grimaces and gestures we make when we are happy or sad, when we are bored, or when we are enjoying ourselves.

MUSCLE TISSUE

All muscles are made of a special tissue that consists of large cells called muscle fibers. When they receive the appropriate order from the nervous system, these fibers contract and the muscle shortens. When the order stops, the fibers relax and the muscle goes back to its original length.

THE MUSCULAR SYSTEM

INTRODUCTION TO THE HUMAN BODY

INDEX

adipose tissue 18, 19
agonist muscle 30
alveoli 12
antagonist muscle 30
aorta 8, 9, 11
arteries 8–9
atrial systole 10
atrioventricular valves 11
atriums 10–11
autonomic nervous
   system 15
axons 14

bacteria 4
bladder 20–21
blood circuits 8
bone system 28–29
brain 14–15
bronchi, left and right
   main 12–13
bronchiole 12

carotid artery 9
cells 4, 5
cerebellum 15
cerebrum 15
cervix 25
circulatory system 6–7,
   8–9, 10–11

dendrites 14
dermis 19
diastole 10
digestive system 16–17

egg cells 24, 25
ejaculatory duct 22
encephalic trunk 15

endocrine system 26–27
endometrium 25
epidermis 19
epididymis 23

fallopian tubes 25
female genitals 24–25
foreskin 22

Galen of Pergamon 7
gonads 26

hair follicles 18
heart 8–9, 10–11
hormones 26–27
hypodermis 19
hypothalamus 26, 27

iliac artery 9
iliac vein 9
inferior vena cava 9
intercostal muscles 13
intestine 16, 17

jugular vein 9

kidneys 20, 21

larynx 13
lungs 12–13

male genitals 22–23
miometrium 25
muscular system 30–31
myocardium 11

nervous system 6, 14–15
neurons 14

ovary 25

pancreas 27
parathyroids 26
penis 22
peripheral nerves 14–15
pharynx 13
pituitary gland 27
prostate 23
protozoa 4
pulmonary arteries 9, 11, 12
pulmonary circuit 8
pulmonary unit 12
pulmonary veins 11, 12

renal artery 9
renal cortex 20
renal medulla 20
renal pelvis 20
renal vein 9
reproductive system
   22–23, 24–25
respiratory system 6,
   12–13

saphenous vein 9
scrotum 22
sebaceous glands 18
seminal vesicle 23
sensory receptors 18, 19
skin 18–19
sperm cell 22, 24
spinal medulla 15
subclavical artery 9
subclavical vein 8
superior vena cava 8
suprarenal glands 27

sweat glands 19
systemic circuit 8

testicle 22, 23
thyroid 26, 27
tibial artery 9
tissues 4, 5
   connective 4, 5
   epithelial 4, 5
   muscle 4, 5, 31
   nerve 4, 5
   osseous 5, 29
trachea 12–13

ureters 21
urethra 21, 23
urinary bladder 21
urinary tract 20–21
urine 20, 21
uterine body 24
uterine cavity 24
uterus 24

vagina 24
vas deferens 22
veins 8–9
vena cava 10
ventricles 10–11
ventricular systole 10